NISTIR 7957

I0467977

Conformance Test Architecture and Test Suite for ANSI/NIST-ITL 1-2011 NIEM XML Encoded Transactions

Fernando L. Podio
Dylan Yaga
Christofer J. McGinnis

Computer Security Division
Information Technology Laboratory

August 2013

U.S. Department of Commerce
Penny Pritzker, Secretary

National Institute of Standards and Technology
Patrick Gallagher, Under Secretary for Standards and Technology and Director

Feedback Suggested

Conformance Test Architectures and Test Suites, User Guides, and sample ("pass/fail") data are available from the following web site:
http://www.nist.gov/itl/csd/biometrics/biocta_download.cfm.

Feedback on these test tools, the sample data, and documentation is welcome. Please send comments to biocts@nist.gov.

Reports on Computer Systems Technology

The Information Technology Laboratory (ITL) at the National Institute of Standards and Technology (NIST) promotes the U.S. economy and public welfare by providing technical leadership for the Nation's measurement and standards infrastructure. ITL develops tests, test methods, reference data, proof of concept implementations, and technical analyses to advance the development and productive use of information technology. ITL's responsibilities include the development of management, administrative, technical, and physical standards and guidelines for the cost-effective security and privacy of other than national security-related information in Federal information systems

Abstract

The latest version of the ANSI/NIST-ITL standard was published in November 2011 (AN-2011). In addition to specifying Record Types in traditional encoding, the standard includes the specification of National Information Exchange Model (NIEM) Extensible Markup Language (XML) encoding and an associated schema. The Computer Security Division of NIST/ITL developed a Conformance Test Architecture (CTA) and Test Suite (CTS) called *BioCTS for AN-2011 NIEM XML* designed to test implementations of AN-2011 NIEM XML encoded transactions. Validating the XML files to a schema may indicate that the contained data is formatted correctly and individual values are within allowable ranges, assuming that the requirements for that data have been documented in the schema file. However, schemas are not designed to test the internal consistency of implementations (i.e., testing for a relationship between two elements or structures within a transaction). These shortcomings of XML schema files for use in conformance testing necessitate that schemas be used only as a component of a complete testing solution. This complete solution (the test tool) ensures test coverage of requirements through a combination of schema validation and conformance tests of the data in the XML files. This document discusses the test software design including the XML Data Structures used and Classes implemented. It addresses the testing phases and the format of the test results; as well as the user interface and key usability features implemented in this version of the test tool. Details are provided on a modified schema that was required to be used in the tool in order to fully perform tests for all the requirements specified in the AN-2011 standard. Future development steps including support for the new version of the ANSI/NIST-ITL standard under development are also discussed.

Keywords

ANSI/NIST-ITL 1-2011, biometrics, conformance testing, conformance test architecture, CTA, CTS, BioCTS, conformance test suite, data interchange formats, encoding, XML NIEM, encoding.

Acknowledgements

The work discussed in this publication was sponsored, in part, by the Department of Homeland Security/Office of Biometric Identity Management (OBIM).

iii

Table of Contents

List of Figures and Tables

1 Introduction

1.1 ANSI/NIST-ITL Standards

The American National Standards Institute/National Institute of Standards and Technology-Information Technology Laboratory (ANSI/NIST-ITL) standard *Data Format for the Interchange of Fingerprint, Facial & Other Biometric Information* is used by law enforcement, intelligence, military, and homeland security organizations throughout the world. The first version of the standard dates to 1986. Over the years, it has been updated and expanded to cover more biometric modalities beyond the original Record Type of fingerprint minutiae. The latest version of this standard was published in 2011. ANSI/NIST-ITL 1-2011 (published as NIST Special Publication 500-290), *Data Format for the Interchange of Fingerprint, Facial & Other Biometric Information* (AN-2011) supersedes all previous versions and amendments to the standard [1]. AN-2011 specifies two data encoding formats: Traditional format (Tag-based format) and a XML format conformant to the National Information Exchange Model (NIEM) [2]. New modalities (DNA and plantars) were added as new Record Types. The extended feature set was added to Record Type 9; Record Type 10 was extended to include all body part images and to include anthropomorphic image markups. Compact iris image storage formats were introduced. There were substantial metadata upgrades as well, including geographic location, data handling logs, original source and associated reference data. Information assurance capabilities were added as Record Type-98.

Development of a new version of the standard started in 2013. A new data record type was introduced for transmission of forensic dental data and imagery. An additional new Record Type was specified to enable transmission of imagery that is not a standard photograph (which would be transmitted in a Type-10 record). Examples include radiographs, Computed Tomography (CT) scans, Positron Emission Tomography (PET) scans, sonograms, 3D orthodontic cast models, Digital Imaging and Communication in Medicine (DICOM) records [3], infrared images, and 3D face data. A new Record Type for Forensic and Investigatory Voice Data has also been specified. The new version of the standard (in draft form at the time of this publication) can be found at the ANSI/NIST-ITL Web Page [4].

1.2 Conformance Testing to Biometric Standards

The existence of biometric standards alone is not enough to demonstrate that products meet the technical requirements specified in the standards. Conformance testing captures the technical description of a specification and measures whether an implementation faithfully implements the specification. Conformance testing provides developers, users, and purchasers with increased levels of confidence in product quality and increases the probability of successful interoperability.

Although no conformance test can be comprehensive enough to test all the different combinations of mandatory requirements of a standard and all possible combinations of conditional and optional characteristics that could be included in a standard, a well-designed conformance test tool that faithfully implements a standard conformance testing methodology could raise the level of confidence on the test results. Therefore, a set of implementations tested with such a tool (and reported to be conformant to the standard), will be more likely to conform to the standard.

1

1.3 NIST/ITL's Computer Security Division Conformity Assessment Related Efforts

The Computer Security Division (CSD) of NIST/ITL supports the development of biometric conformance testing methodology standards and other conformity assessment efforts through active technical participation in the development of biometric standards and associated conformance test architectures and test suites. NIST/ITL CSD develops these test tools to support users who require conformance to selected biometric standards and product developers interested in conforming to biometric standards by using the same testing tools available to users. Testing laboratories can also benefit from the use of these test tools. These efforts support the possible establishment of conformity assessment programs to validate conformance to biometric standards.

Under conformance test software called "BioCTS", NIST/ITL CSD has developed a number of Conformance Test Architectures (CTAs) and Conformance Test Suites (CTSs) to test implementations of national and international biometric data interchange formats. The biometric conformance testing software includes the test tools designed to test implementations of the following standards:

- Previous and current versions of the ANSI/NIST-ITL standards (1-2007 and 1-2011 versions) [4] for the Traditional (tag-based) encoding.

- Biometric data interchange formats specified in binary encoding developed by the International Committee for Information Technology Standards Technical Committee M1- *Biometrics* (INCITS/M1) [5].

- Biometric data interchange formats specified in binary encoding developed by ISO/IEC Joint Technical Committee 1 Subcommittee 37 – *Biometrics* (JTC 1/SC 37) [6].

- Selected PIV Profiles[1] specified in NIST Special Publication 800-76-2, *Biometric Specifications for Personal Identity Verification* published in July 2013 [7].

These tools, sample data (including passing data files and failing data files) and associated documentation are available to the public and can be downloaded from the NIST/ITL CSD test tool download web site [8]. Previous work from NIST/ITL CSD, such as NIST/ITL's Biometric Application Programming Interface (BioAPI) Conformance Test Suite and NIST/ITL's Conformance Test Suite for Patron Format A Data Structures Specified in ANSI INCITS 398-2008, Common Biometric Exchange Formats Framework (CBEFF) are available at [9] and [10] respectively.

The NIST/ITL CSD download web site has the latest test tool releases which include an upgraded version of the CTAs to support the CTSs listed above as well as new CTSs for additional international standards and a CTS to test AN-2011 NIEM XML encoded transaction which will be discussed in the next sections. Plans include migrating the latest ANSI/NIST-ITL test tool to support the 2013 new version and new CTS designed to test XML encoded international data interchange formats.

---

[1] PIV Profiles of ISO/IEC 19794-6:2011 (On Card/Off Card), ANSI/INCITS 378:2004, and ANSI/INCITS 381:2004.

2.1 Structure of AN-2011 XML-Encoded Transactions

The AN-2011 NIEM-conformant XML encoded transactions are specified within an XML Exchange Package; which uses specifically named XML Elements to represent data structures in a strictly ordered fashion. The `<itl:NISTBiometricInformationExchangePackage>` XML Element must be the root Element, and its children must be Elements associated with the Record Types defined in the Traditional encoding. All information regarding the complete structure of an Exchange Package can be found in Annex C and Annex G of the AN-2011 standard and the associated XML schema [11] that can be found at the ANSI/NIST-ITL Standard Web page. As discussed in Section 2.4 below, the specified schema does not enforce all of the requirements of the AN-2011 standard and therefore, the test tool relies on a modified schema ("default" schema).

The AN-2011 standard defines the structure of AN-2011 traditionally encoded Transactions as being comprised of Records, Fields, Subfields, and Information Items. In NIEM-XML encoded Transactions, known as Exchange Packages, named XML Elements represent all data structures defined in the AN-2011 standard. The mapping of these named XML Elements to the structures referenced in the AN-2011 standard is specified in AN-2011 Annex G: *Mapping to the NIEM IEPD*. The AN-2011 standard defines requirements for both encodings, but references structures using their Traditional encoding identifiers such as Record Types, Field Numbers, and Subfield and Information Item positions. These requirements must be translated to the XML equivalent using the mapping provided in Annex G. Some specific issues, including differences between the two encodings, must be taken into consideration when performing this translation:

- **Name Uniqueness:** The named XML Elements are not necessarily unique, meaning that one XML Element name may be used to represent one or more structures. As a result of this difference, the name of the XML Element and its relative position to other Elements are important in identifying the appropriate Element within the Exchange Package. For example, querying the Exchange Package for `<biom:RecordCategoryCode>` will not return a single specific structure, because that Element name represents Field xx.001 in every Record Type.

- **Ordering:** While Traditional encoding requires only some structures to be ordered, such as the first two Fields of each record, XML encoding requires strict ordering of Elements.

- **Optional Structures:** In Traditional encoding the information separators for optional Information Items must always be present. Optional XML Elements are simply omitted from the Exchange Package when not used.

- **Mapping Cardinality**: There is not always a one-to-one relationship between XML Elements and Traditional structures. Some examples of differences include:

  - **There is no XML equivalent for the Traditional structure**. Such as the case when a field contains subfields in Traditional, but the XML instead uses Elements to represent the subfields directly, without the need for a Field to contain them.

- There is no Traditional equivalent for the XML Element. Most often this occurs when an XML Element is used to contain or group several sets of related XML Elements, but this grouping does not exist in Traditional encoding.

- Two or more XML Elements are used to represent one Traditional structure. Most often this occurs when multiple XML Elements are used to categorize ranges of allowed data within a single Traditional structure using multiple Elements. For example, `<biom:FaceImageDescriptionCode>` and `<biom:FaceImageDescriptionText>` both map to Field 10.026.

- **NIEM-defined Elements**: In an effort to conform to NIEM encoding rules, the Exchange Package makes use of a standard set of XML Elements to represent common data structures. These Elements are not listed in the mapping, and are only mentioned in the XML schema. Some examples include: `<nc:DateTime>`, `<nc:Date>`, and `<nc:IdentificationID>`.

2.2 *BioCTS for AN-2011 NIEM XML* Design

BioCTS for AN-2011 NIEM XML, developed by NIST/ITL CSD, is a CTA/CTS designed to test implementations of NIEM XML encoded transactions for the requirements specified in AN-2011. The conformance software tool implements all test assertions developed by NIST/ITL CSD and documented in NIST IR 7806 *ANSI/NIST-ITL 1-2011 Requirements and Conformance Test Assertions* [12] based on the requirements specified in AN-2011.

NISTIR 7806 documents over 1,200 conformance test assertions for selected AN-2011 Record Types. They are divided into tables of requirements of assertions for each Record Type. The assertions are identified as being applicable to the Traditional encoding only, the NIEM-XML encoding only, or both Traditional and NIEM XML encoding. These tables of requirements and the associated test assertion syntax were later adopted as part of the conformance testing methodology documented in NIST Special Publication 500-295, *Conformance Testing Methodology for ANSI/NIST-ITL 1-2011, Data Format for the Interchange of Fingerprint, Facial & Other Biometric Information* [13].

In addition to the NIEM XML encoding test assertions documented in NIST SP 500-295, some modified assertions and new updated assertions developed after the publication of NIST SP 500-295 were developed in code. A revision of NIST SP 500-295 is planned reflecting these changes. These test assertions are designed to test AN-2011 transactions that include selected record types of AN-2011[2]. Transactions that include other record types are also tested for consistency to the standard. Exceptions to the test assertion implementations and their rationale are detailed in Annex A.

---

[2] The supported AN-2011 sessions and Record Types (RT) include Section 5 (Data Conventions) and 7: (Information Associated with Several Record Types), RT-1: Transaction information record, RT-4: Grayscale fingerprint image, RT-10: Facial, other body part and SMT image record, RT-13: Friction-ridge latent image record, RT-14: Fingerprint image record, RT-15: Palm print image record, RT-17: Iris image record, Annex B: Traditional Encoding and Annex C: NIEM-conformant encoding rules.

The design of the BioCTS for AN-2011NIEM XML CTA/CTS was a result of careful analysis of the base requirements in the ANSI/NIST-ITL 1-2011 standard, the NIEM-Encoding specific requirements, and the NIEM XML schema definitions. The design reflects the fact that using the NIEM schema alone is not enough to fully test for conformance to the base requirements specified in the standard – the schema alone can ensure that the data is formatted correctly, but not ensure if it is conforming to all of the base requirements.

2.3 Test Phases

BioCTS for AN-2011 NIEM XML has three test phases:

1. **Determination of whether the XML file is Well-Formed.**
   An XML file is Well-Formed when it is syntactically correct, and follows the rules of XML documents [14], [15]. Without a Well-Formed XML file, further testing would yield potentially unusable results.

   *If the XML file fails this portion of testing, it **will not continue** to the next phase.*

2. **Validating the XML file against the specified schema file.**
   BioCTS for AN-2011 NIEM XML will attempt to validate the XML file against the specified schema file (the test tool allows for a user-defined schema file, but also provides a default), and will report as many errors as possible.

   *The test tool, regardless of whether the XML file passes or fails this phase of testing, **will continue** to the next phase.*

3. **Assertion testing for AN-2011 base requirements**

   The final phase is testing against the base requirements of ANSI/NIST-ITL 1-2011 that are not covered by the XML schema file validation. These tests include:

   a. Valid Value tests

   b. Character Count tests

   c. Relationship tests between XML elements

   d. Basic Image Validation

After all three phases of testing are complete for the XML file, the test results are aggregated and an overall result (Pass or Fail) is determined. For the overall result of an XML file to be reported as "Pass" there must be no result of "Error" or "Critical" in any of the XML file's results.

Fig. 2-1 depicts the test phases performed within the test tool to determine the conformance of an AN-2011 NIEM XML implementation.

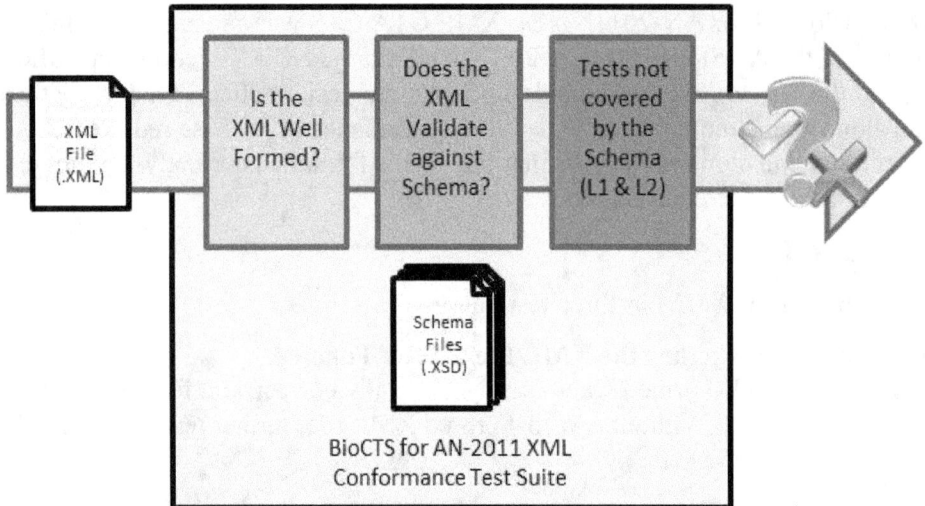

Figure 2-1: *BioCTS for AN-2011 NIEM XML* **Test Phases**

2.4 Schema Modifications Required to Perform the Tests

Section C.5.1 of the AN-2011 standard specifies:

> *To the extent possible, the schema define data types and constraints that enforce the allowable content rules of the base standard. Nevertheless, the XML schema may not strictly enforce the allowable content. The base standard defines allowable content, and its requirements shall be met by implementers regardless of encoding method.*

Based on the premise that the conformance test tool needs to be designed to perform tests on implementations of the "base standard" as stated above, the tool uses a modified schema (which is used as the defaults schema). A comprehensive comparison of AN-2011 requirements and the Schema files provided [16] led to the identification of the following types of discrepancies:

- The schema is **LESS** restrictive than the base requirements, and may allow additional values other than what is defined in the base requirements.
    - *In this case, the schema **does not** prevent an implementation from passing the tests.*
    - Additional tests have been implemented in the *BioCTS for AN-2011 NIEM XML* Encoded Transactions software to ensure that the values found within an XML encoded transaction conform to the base requirements.

- The schema is **MORE** restrictive than the base requirements and may not allow certain values that are allowed in the base requirements.
    - *In this case, the schema **does** prevent an implementation of the AN-2011 standard from passing the tests.*
    - The BioCTS team modified the schema files distributed with the AN-2011 standard.
    - Additions and modifications were made to the Schema files in the cases where the Schema prevented the base requirements from passing.

6

○　These modifications are reflected in the Schema file included with the *BioCTS for AN-2011 NIEM XML* software.

A summary of the required Schema modifications can be found in Annex B.

2.5 *BioCTS for AN-2011 NIEM XML* Data Structures and Class Diagrams

ANSI/NIST-ITL 1-2011 Exchange Packages include named XML Elements that map to Records, Fields, Subfields, and Information Items defined in the Traditional encoding. As stated above, the mapping is not necessarily one-to-one, and specific Elements required by the NIEM encoding rules (and not related to any structures in Traditional Encoding) are used to represent data in some cases. BioCTS for AN-2011 XML makes use of Inheritance and Polymorphism to define the XML Elements in the Exchange Package, as well as to indicate their relationship to identifiers used in the AN-2011 standard, such as Record Types, Field Numbers, and Mnemonics. For example, Abstract Classes are used when possible to define functionality for a set of related structures and to ensure that only derived types of those Classes are instantiated. This helps ensure that only well-defined structures are parsed and tested, and that all unrecognized structures are reported correctly. In addition, the derived Classes can all be referenced as the base Class or specifically selected based upon their types.

2.5.1 Auxiliary Classes

As shown in Fig. 2-2, Several Auxiliary Classes are defined in the testing architecture.

These Auxiliary Classes are used throughout the software to assist in the overall process of parsing, searching for and testing of the XML Elements found in an Exchange Package. The Xml_Information Class, which includes the Xml_Namespace Class and Element name, is used throughout BioCTS to reference specific XML Elements. The XmlEncoding Class is used to ensure the XML file uses one of the supported encoding types (UTF-8, UTF-16, and UTF-32).

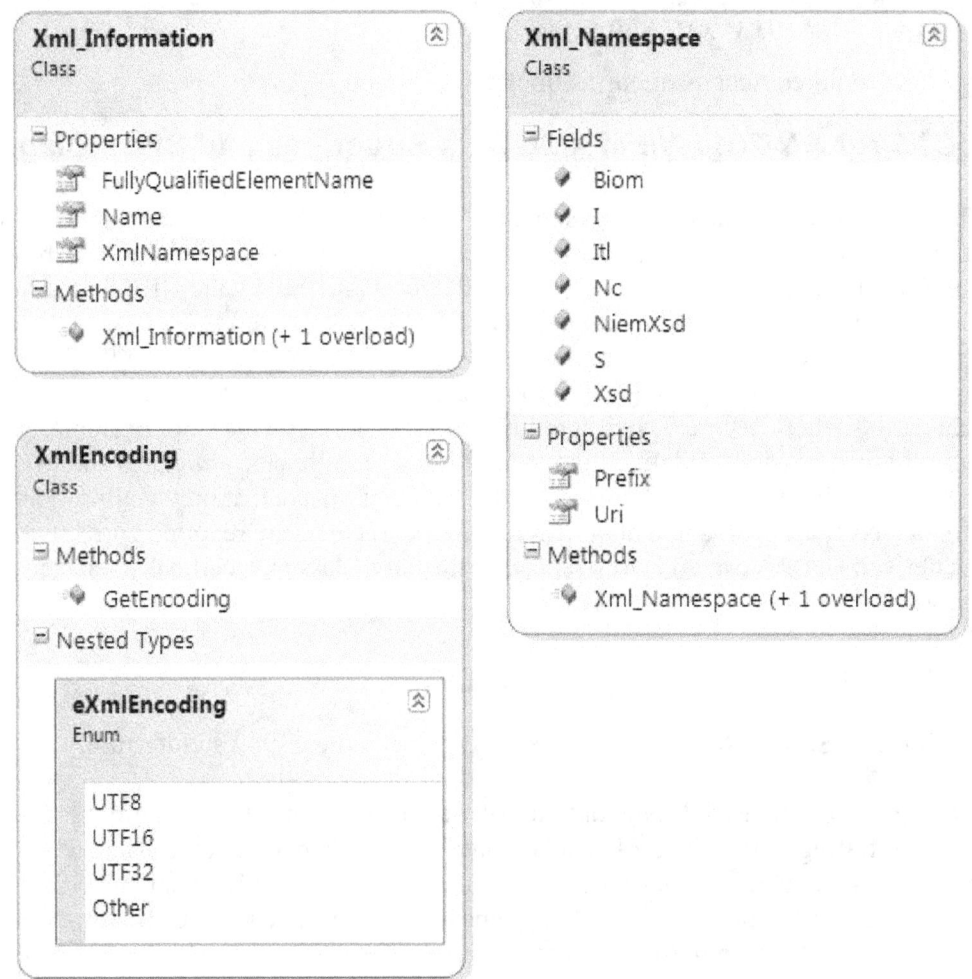

Figure 2-2: Auxiliary Classes

2.5.2 Class: An2K11_Xml

The An2K11_Xml Class represents the XML Exchange Package and its metadata, such as the Encoding and XML Validation Information. An2K11_Xml also initializes several utility classes used for parsing the XML document. It contains a reference to the Xml_Transaction Class, which represents the root XML Node of the Transaction. Fig. 2-3 depicts the An2K11_XML Class.

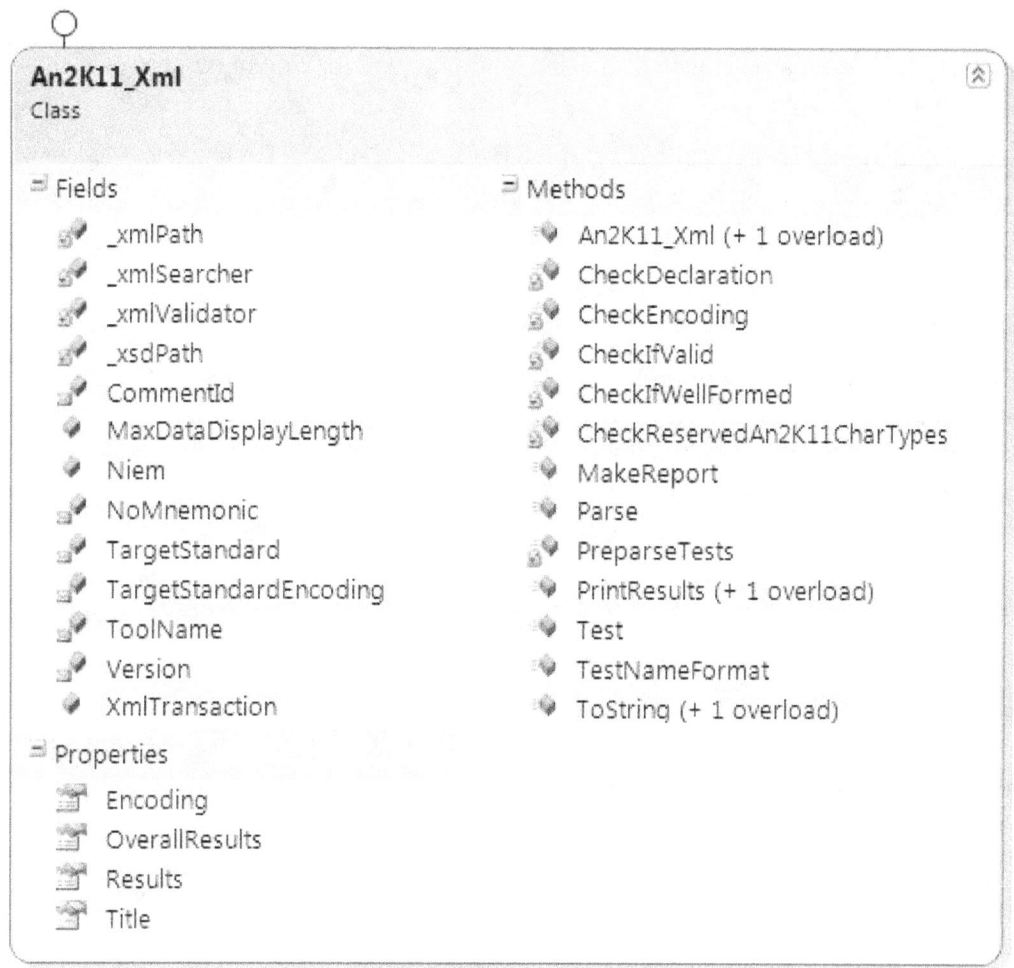

Figure 2-3: An2K11_XML Class

The `An2K11_Xml` Class plays an important role in testing as well, because it performs any tests that are not directly associated with a specific XML Element, such as ensuring that the XML element follows the rules and syntax of XML specifications [17], XML schema Validation, and AN-2011 reserved character checks.

2.5.3   Abstract Class: Element

BioCTS for AN-2011 NIEM XML treats all structures in an Exchange Package as an `Element`. `Element` is a top-level Abstract Class from which all other XML Element categories are derived. Figure 2-4 illustrates the class hierarchy for `Element` types and provides a class diagram for `Element`. The `Element` class provides much of the common functionality that all XML Elements must have. Each `Element` has:

- **Cardinality indicators:** `CardinalityMin` and `CardinalityMax` as defined in Annex G of the AN-2011 standard. This indicates the minimum and maximum number of occurrences for the `Element`.

- **Elements:** Due to the hierarchical nature of the Exchange Package, each `Element` contains a list of child `Elements`, starting from the top with the `Xml_Transaction Element`.

- **Mandatory Elements:** A list of `Elements` that must be present. These `Elements` have a `CardinalityMin` of 1 or higher and are listed in the standard as having a `CondCode` of Mandatory (see Section 8 of AN-2011) – NIST SP 500-290.

- **Results:** The `OverallResults` for each `Element` is used to indicate whether any of the tests related to that Element or any of its child Elements have failed. If the `OverallResults` Pass, then all tests for that `Element` and its child `Elements` have passed. Note that in Figure 2-4, `Element` implements the `ITestable` Interface, which sets up the contract for classes that should report `OverallResults`.

- **RootNode:** BioCTS uses the `XmlNode` class build into the C# language to contain the actual XML Element data and position within the XML document. The `RootNode` field holds the `XmlNode` reference to the `Element` that the `Element` instance is representing.

- **Formatting:** `TestTitle` is used to indicate the way the `Element` should be listed in test names as define in NIST SP 500-295 (AN-2011 CTM), and `Title` indicates how the `Element` should appear for all other uses.

Each Element also has a set of methods available, and already defined by the base class, including:

- **Get Elements:** The `GetElement()` methods allow each `Element` to be queried for specific `Elements` by name, type, or both. Type is any derived `Element` type shown in Fig. 2-4, and Name is an `Xml_Information` type, which is a Class that includes the Xml namespace and `Element` name. The search can be performed on direct children only or recursively.

- **Occurrence Errors:** The `GetCardinalityErrors()` method provides a formatted string indicating which child `Elements` have invalid cardinality (their occurrence is outside of the range of acceptable values). The `GetMissingMandatoryElements()` specifically returns a formatted string indicating which mandatory child `Elements` are missing.

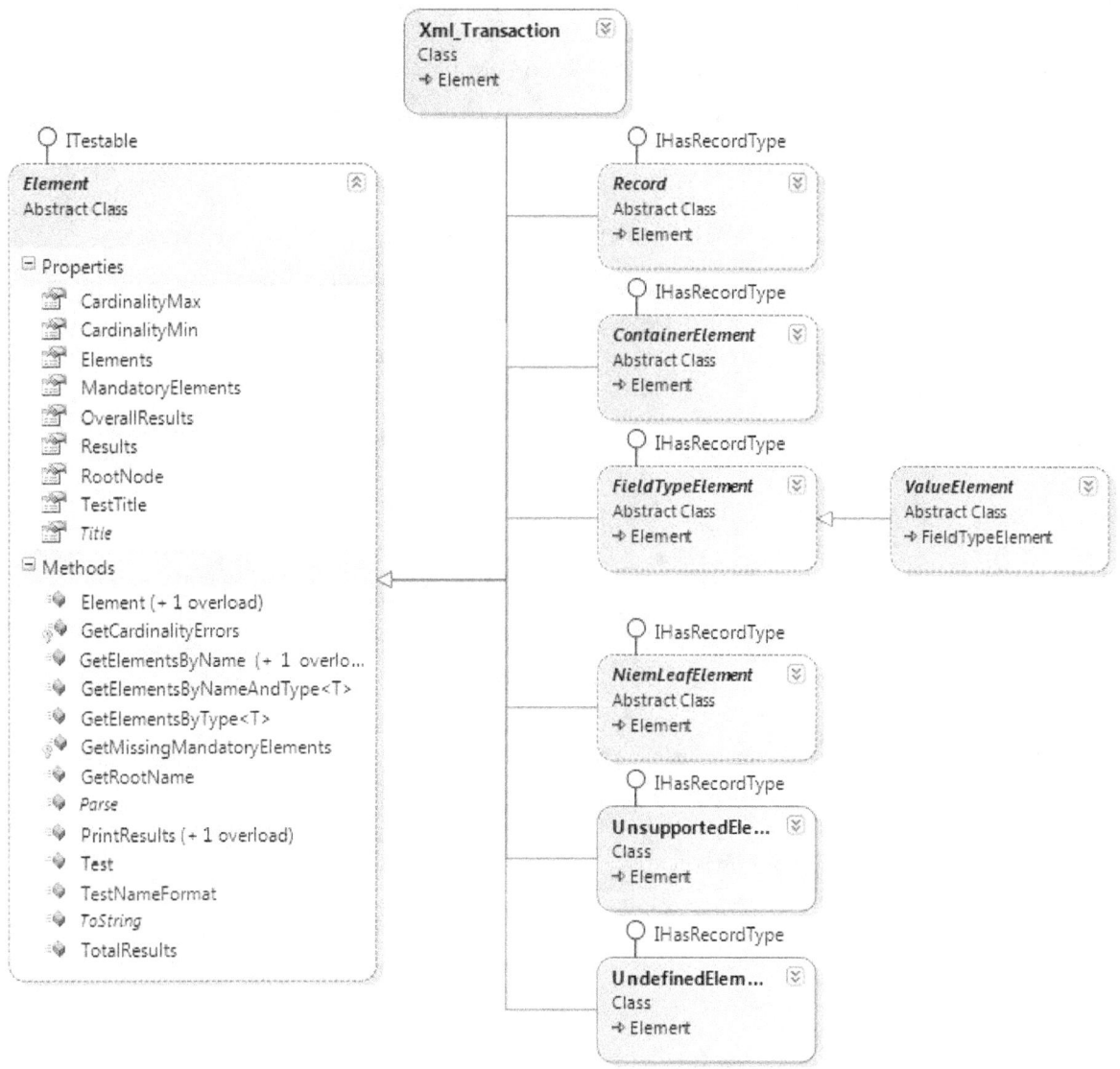

Figure 2-4: "Element" Class and Inheritance

The classes that inherit from the `Element` Abstract Class are used to categorize the types of `Elements` in an Exchange Package. While all of these `Elements` are simply XML Elements, they do loosely map to some of the structures defined in the base AN-2011 standard and Traditional encoding.

2.5.4 Class: Xml_Transaction

This Class represents and contains the Root Node of the Exchange Package:
`<itl:NISTBiometricInformationExchangePackage>`. It contains a list of all Record
Elements and provides methods for accessing the Records.

All Transaction-level assertion tests, as specified by the AN-2011 CTM, are defined in this class.

2.5.5 Abstract Class: Record

The `Record` Class represents any XML Element that is associated with a Record Type.

While the `Record` Class is Abstract and cannot be instantiated, it provides fields and methods that
are common among all AN-2011 Record Types, as shown by the Class Diagram in Fig. 2-5. The
classes that inherit from `Record` are the various Record Types that represent different modalities in
the AN-2011 standard.

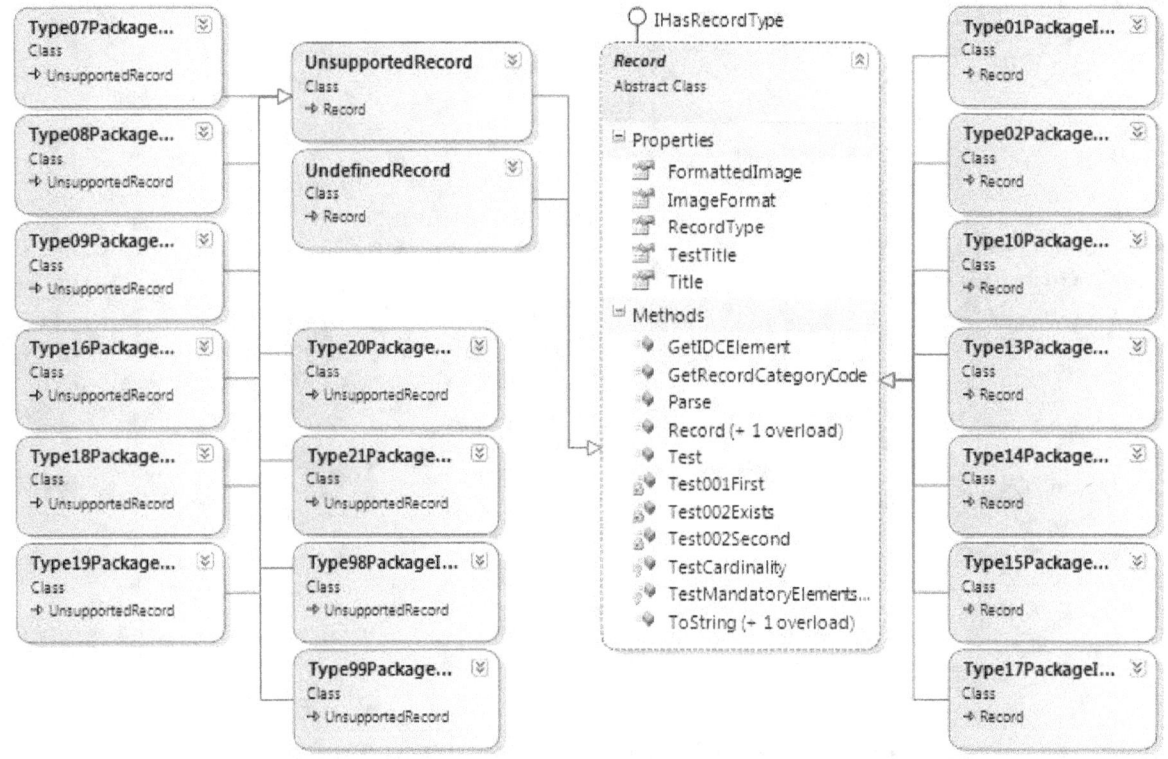

Figure 2-5: "Record" Class and Inheritance

In addition to the supported Record Types, there are two additional classes that inherit from
`Record`.

- **UndefinedRecord:** This class is used to contain any XML Element that is a child of the
  `Xml_Transaction` Element (the Exchange Package root node) but is not defined as a
  valid Record Type. *These Records are reported as errors by BioCTS.*

- **UnsupportedRecord:** This class is used to contain any XML Element that represents a valid AN-2011 Record that is not fully supported by BioCTS. While these records remain available for conformance testing, they are not fully tested. In addition, each of these records produces a warning indicating that the Record is only partially supported.

2.5.6 Abstract Class: ContainerElement

The `ContainerElement` Class is used to represent any XML Element that is not directly related to a structure defined in AN-2011 (such as Fields and Information Items), but rather indicates a grouping of related Elements. Table 2-1 displays several Container Elements including `<biom:FaceImage>`, `<biom:PhysicalFeatureImage>`, and `<biom:ImageCaptureDetail>` among others.

Field ID	Mnemonic	XML element name	Cardinality
10.997	SOR	biom:SourceRepresentation	0..255
"	SRN	biom:SourceIdentification	1..1
"	RSP	biom:ImageSegmentIdentification	0..1
-	-	biom:FaceImage[96]	0..1[97]
-	-	biom:PhysicalFeatureImage[98]	0..1
10.999	DATA	nc:BinaryBase64Object	1..1
-	-	biom:ImageCaptureDetail	1..1
10.998	GEO	biom:CaptureLocation	0..1
"	GRT	nc:LocationDescriptionText	0..1
"	-	nc:LocationGeographicElevation	0..1
"	ELE	nc:MeasurePointValue	1..1
"	-	biom:LocationTwoDimensionalGeographicCoordinate	0..1
"	-	nc:GeographicCoordinateLatitude	0..1
"	LTD	nc:LatitudeDegreeValue	0..1
"	LTM	nc:LatitudeMinuteValue	0..1

Table 2-1: ContainerElement Examples

2.5.7 Abstract Class: FieldTypeElement

The `FieldTypeElement` Class is used to represent any XML Element that is directly mapped to a Field defined in AN-2011 Annex G. It extends the `Element` Class by adding the `FieldNumber` and `Mnemonic` properties.

2.5.8 Abstract Class: ValueElement

The `ValueElement` Class inherits from the `FieldTypeElement` Class, and is used to represent any XML Element that is directly mapped to any structure in AN-2011 that contains data. The ValueElement Class Diagram is depicted in Fig. 2-6.

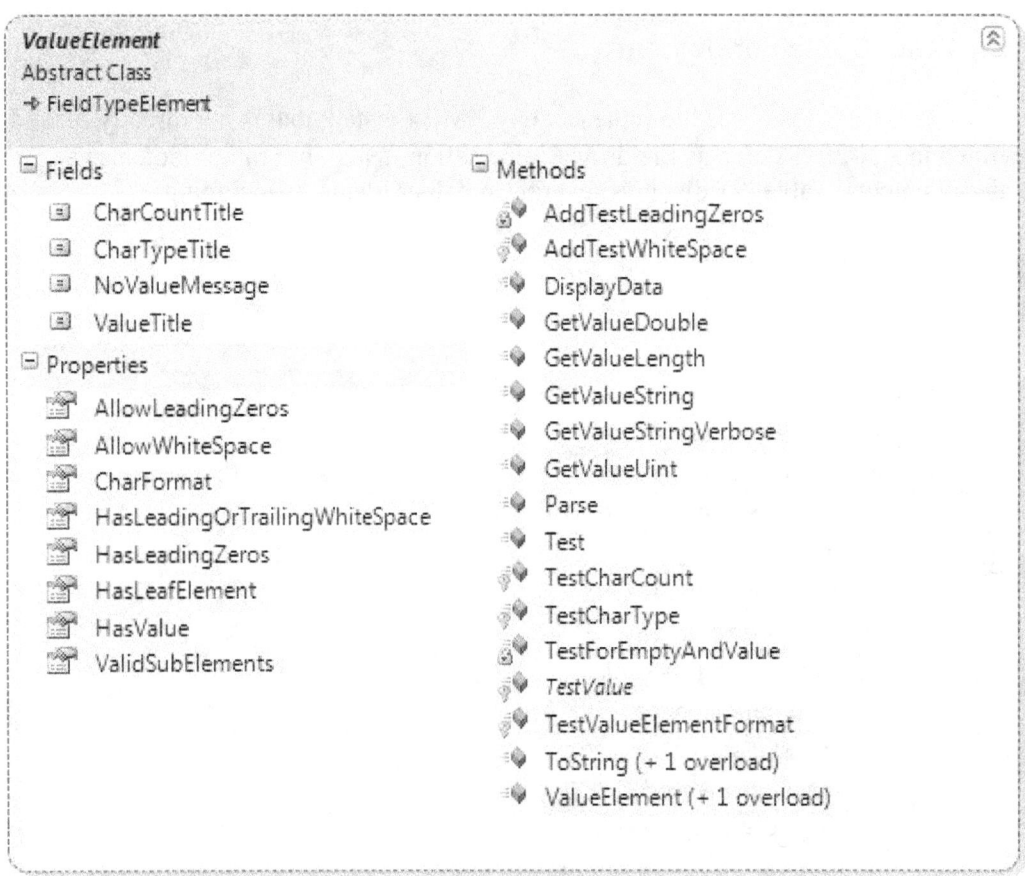

Figure 2-6: ValueElement Class Diagram

This is an important component of the BioCTS design that is specific to the XML encoding. While the Traditional encoding specifies several types of containers for data including Fields, Subfields, and Information Items; *BioCTS for AN-2011 NIEM XML* considers all of these structures to be `ValueElements` in the XML encoding. This allows more flexibility in the parsing, searching, and reporting of entities that contain data. Because the majority of test assertions are concerned with the data contained in structures, the `ValueElement` class is a vital and extremely useful piece of the BioCTS XML architecture.

Fig. 2-6 depicts the methods and properties that the ValueElement Class provides, some of which are described below:

14

- **Whitespace Indicators:** `AllowWhiteSpace` and `AllowLeadingZeros` are used to indicate which fields should allow whitespace and leading zeros and which should not. If whitespace is detected when it is not specifically allowed, a warning is displayed and the whitespace is ignored for further tests.

- **Characters Allowed:** `CharFormat` is used to specify the exact `CharType` requirements found in the AN-2011 standard (see Section 5.5 of AN-2011).

- **Get Data:** `GetValueString`, `GetValueLength`, `GetValueDouble`, and `GetValueUint` allow the retrieval of data from the Element in the desired format.

- **Leaf Elements:** `HasLeafElement` and `ValidSubElements` are used to indicate whether or not NIEM-specific Elements are supposed to be contained within the `ValueElement` to contain the data. These Elements, when present, are defined only in the XML schema and not in Annex G of the AN-2011 standard.

2.5.9 Abstract Class: NiemLeafElement

The `NiemLeafElement` Class is used to define NIEM-specific Elements required by NIEM encoding rules to contain common data types. They are found within the `ValueElements`. These Elements, when present, are defined only in the XML schema and not in Annex G of the AN-2011 standard. These Elements are not reported in the test results other than as a Sub-Element to a `ValueElement`.

2.5.10 Abstract Class: UnsupportedElement

The `UnsupportedElement` Class is used to contain XML Elements that are defined by AN-2011 but not yet supported by BioCTS. They are reported as warnings during testing.

2.5.11 Abstract Class: UndefinedElement

The `UndefinedElement` Class is used to contain XML Elements that are not defined by AN-2011. They are reported as errors during testing.

2.6 Parsing

BioCTS for AN-2011 NIEM XML has added levels of complexity when parsing XML Transactions, as opposed to Traditional Transactions. Parsing can be broken down into specific levels: Pre-Parse, `Xml_Transaction Parse()` and Hierarchical Parsing. Each stage of parsing is critical and must complete without errors for BioCTS to be able to successfully accept and test XML Encoded Transactions. Failing at the parsing level will cause BioCTS to stop testing a XML Transaction, as any Results derived from an unsuccessful/incomplete parse could not be considered reliable.

2.6.1 Pre-Parse

The following major operations occur during Pre-Parsing.

(a) Determine if the File Path for the Implementation Under Test passed into the software exists (e.g., a valid file). If not, the file is skipped and no results are reported.

(b) Determine if the XML schema File Path passed into the software exists. If not, the default XML schema is used.

(c) Check for valid AN-2011 characters. If invalid characters are found, they are reported as an error, and then the parsing and testing proceeds ignoring those characters.

(d) Check for the root node `<itl:NISTBiometricInformationExchangePackage>`. If the root Element is not found, an error is reported and *BioCTS does not perform any further processing.*

2.6.2 Parse Elements

If the Pre-Parse tests pass successfully, the `Xml_Transaction` Class is instantiated with the root node of the Exchange Package and the `Xml_Transaction Parse()` method is called to begin the parsing process.

  o *Xml_Transaction Parse()*

This Transaction-level `Parse()` method calls the `Refine()` method for each child Element of the Transaction's root, which checks the Element name for valid Record names as defined in Annex G of AN-2011. All valid Records are added to `Xml_Transaction` Element list. Any invalid Elements are added to the `Element` list as `UndefinedRecord` types, which are reported as errors.

The `Xml_Parse()` method then calls the `Parse()` method for each `Record` added to the `Element` list.

  o *Hierarchical Parsing*

This parsing process continues to propagate down the class hierarchy, as each Element refines the XML Elements found in its root node by comparing the Element names to valid names found in Annex G of AN-2011. Then, each Element iteratively calls each of its child Elements' `Parse()` methods and the process repeats. When invalid XML Elements are discovered, they are added as `UndefinedElement` types, and reported as errors. Note that the parsing does not take order into consideration. The NIEM-XML encoding requires strict ordering of Elements; however, the ordering is not tested during parsing. Element ordering is reported during XML schema validation.

2.7 Testing

Testing can begin only after parsing is successfully completed. Each test has been designed to require no more information than necessary. As a result, several "Testing Levels" can be observed depending on how much information is required for a particular test (from most information needed → least):

- **An2k11_Xml Testing**: Requires the XML document file path and the XML schema file path

- **Transaction-Level Testing**: Needs information from more than one `Record`

- **Record-Level Testing**: Needs information from more than one `Element` within the same `Record`

- **Element-Level Testing**: Needs no more information than what is stored within a single `Element`

2.7.1 An2k11_Xml Testing – (`An2Kk11_Xml.Test()`)

This level of testing is performed at the Exchange Package metadata level and includes tests for ensuring that the XML file is well formed and validates against the schema, and that the XML declaration is present, along with the XML encoding. These tests are not related to any of the Elements contained in the Exchange Package or any of the structures defined in AN-2011. If the well-formed test fails, testing does not proceed.

After these tests are performed, the `Xml_Transaction Test()` method is called.

2.7.2 Hierarchical Element Testing

All Classes that derive from the Element class make use of the `Element.Test()` method shown below. This simple, iterative approach allows each Class that inherits from Element to override the `Element.Test()` method, add its own tests, and then call `base.Test()` in order to continue the testing process for each of its child Elements.

```
public virtual void Test()
{
    foreach (var elem in Elements)
    {
        elem.Test();
    }
}
```

To illustrate this approach, the following pseudo code shows how the `Xml_Transaction` performs its tests and then uses the `Element.Test()` method to continue testing each of its children. This process continues down the inheritance tree until all Elements have been tested.

```
public override void Test()
{
    //Level 1 Transaction Tests

    //Level 2 Transaction Tests

    //Call Element.Test()
    base.Test();
}
```

3   Conformance Test Architecture User Interface Features

As shown in Fig. 3-1, the latest Conformance Test Architecture released in August 2013 supports Conformance Test Suites designed to test transactions encoded in Traditional encoding (Tag-Based) and XML encoding (NIEM XML). Single-file testing or Batch Testing (loading several Transactions (files) at once) can be performed.

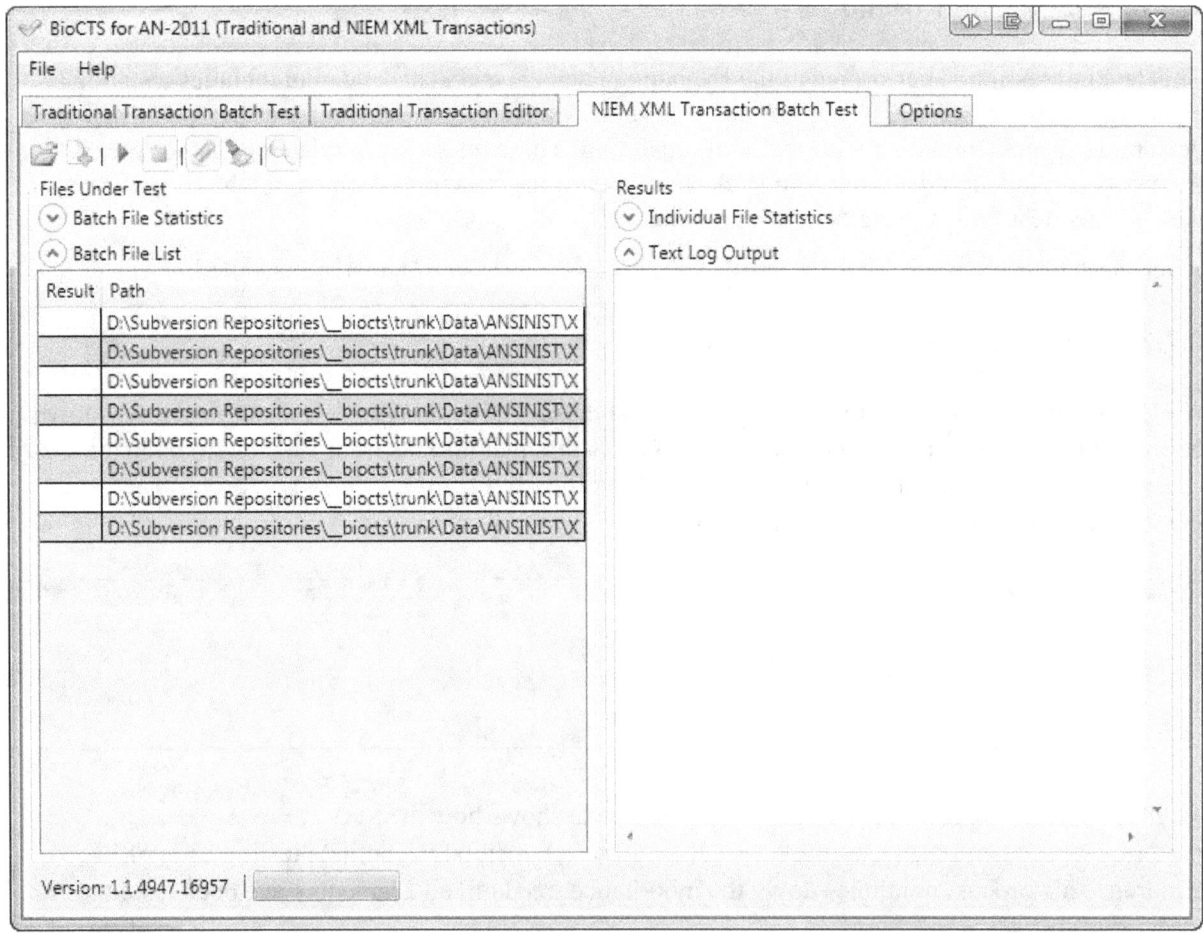

Figure 3-1 – Support for Tag-Based and NIEM XML Encoded Transaction Testing

An Editor for Tag-Based Transaction editing is included. There are many tools (both freely and commercially available) that allow for editing of the Text-based XML files. Currently, *BioCTS for AN-2011 NIEM XML* Encoded Transactions does not include an editor for these Transactions. An "Options" User Interface provides options for output file type, constraint schema path and Text and XML Log Output save locations. More details on the available options are included below.

New Key Features included in this version are Text Log Output Search and Test Result Statistics.

- *Text Log Output Search*

Searching through an entire Text Log can be difficult, especially when a Transaction has multiple Records within it. To help alleviate this problem, without forcing the user to open an external tool, BioCTS for AN-2011 has implemented an internal search feature. To launch the search window, the magnifying glass icon can be clicked or the keyboard command CTRL+F can be used. This feature is shown in Fig. 3-2.

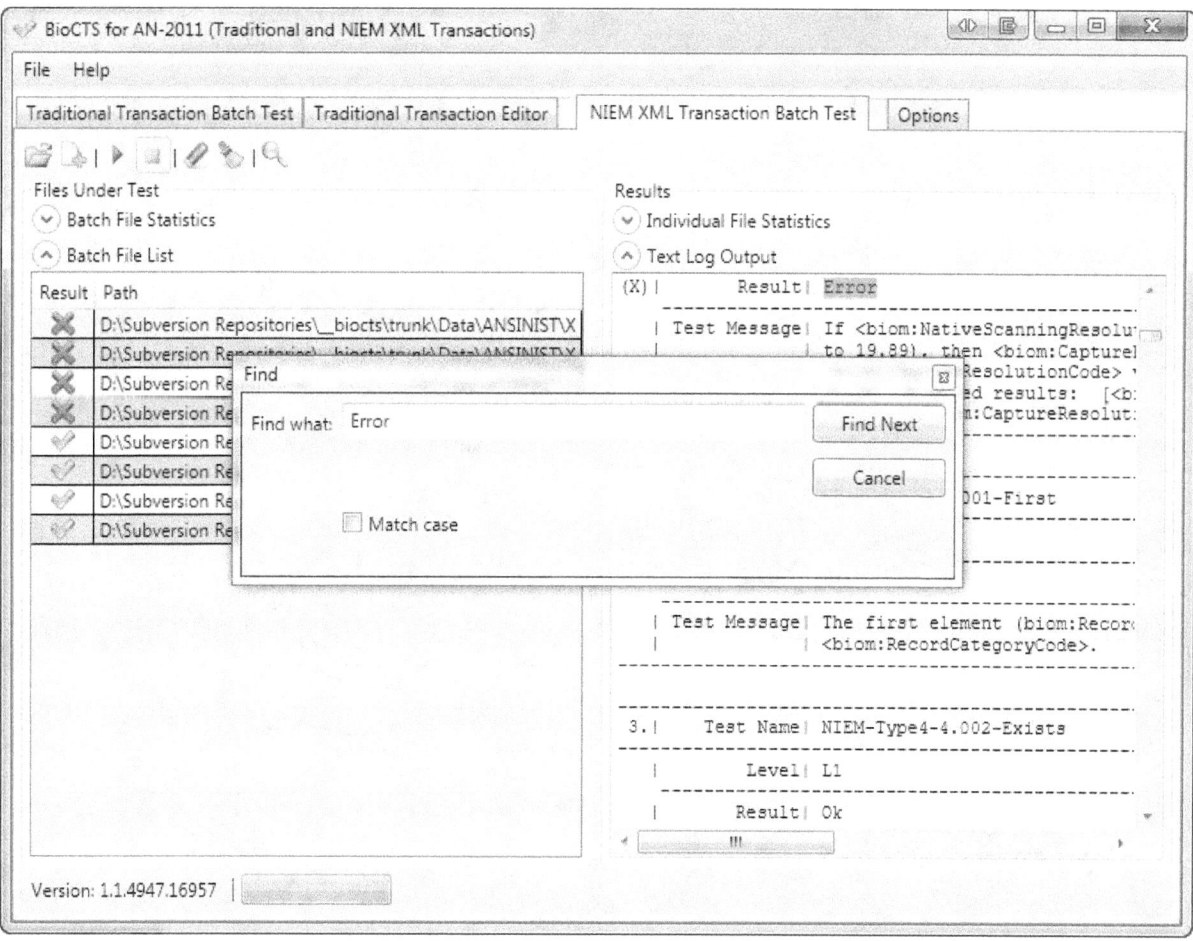

Figure 3-2 - BioCTS Text Log Output Search Feature

- *Statistics*

In previous releases of BioCTS it was difficult to quantify many test metrics such as the number of files loaded and tested, the number of tests performed, and how many tests resulted in an error or warning. Basic statistics are now available including:

- o **Selected File Statistics**: Details about how many tests were performed and a breakdown of what types of results were found. When a file is selected within the Batch XML Test tab, the statistics are updated to reflect that file.

- o **Overall Batch Statistics**: Details about how many files were loaded into the Batch File tester, how many files were tested, and how many files were considered "Passing" or "Failing".

Fig. 3-3 highlights the statistical summaries.

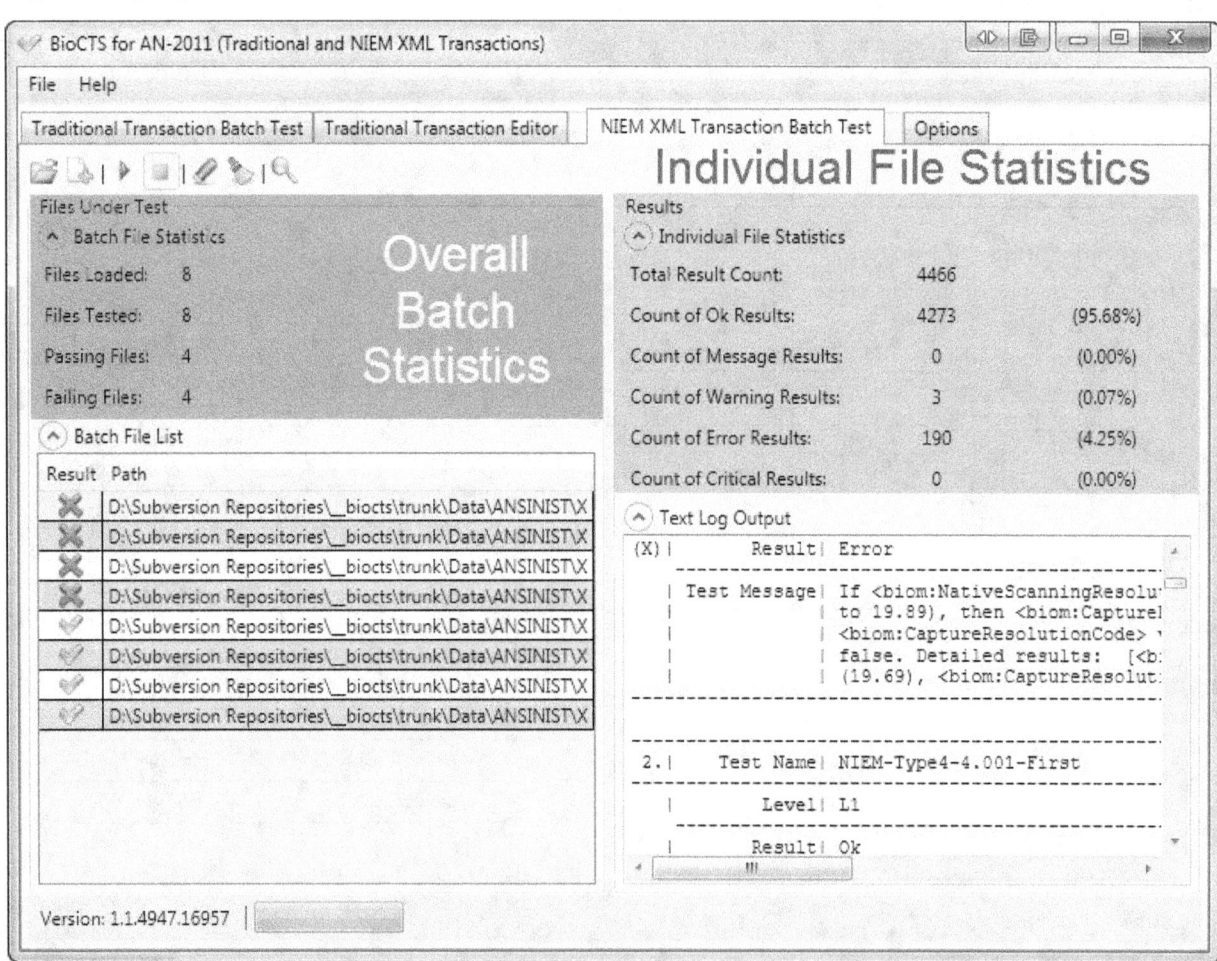

Figure 3-3 - BioCTS Highlighting the Statistic Summaries

3.2.1 Batch Testing

As shown in Fig. 3-4 and discussed above, many Transactions (files) can be loaded at once and tested in groups. The "Batch Test" tab will display the transaction's overall result with either:

✖ - Overall Result of Fail / ✔ - Overall Result of Pass

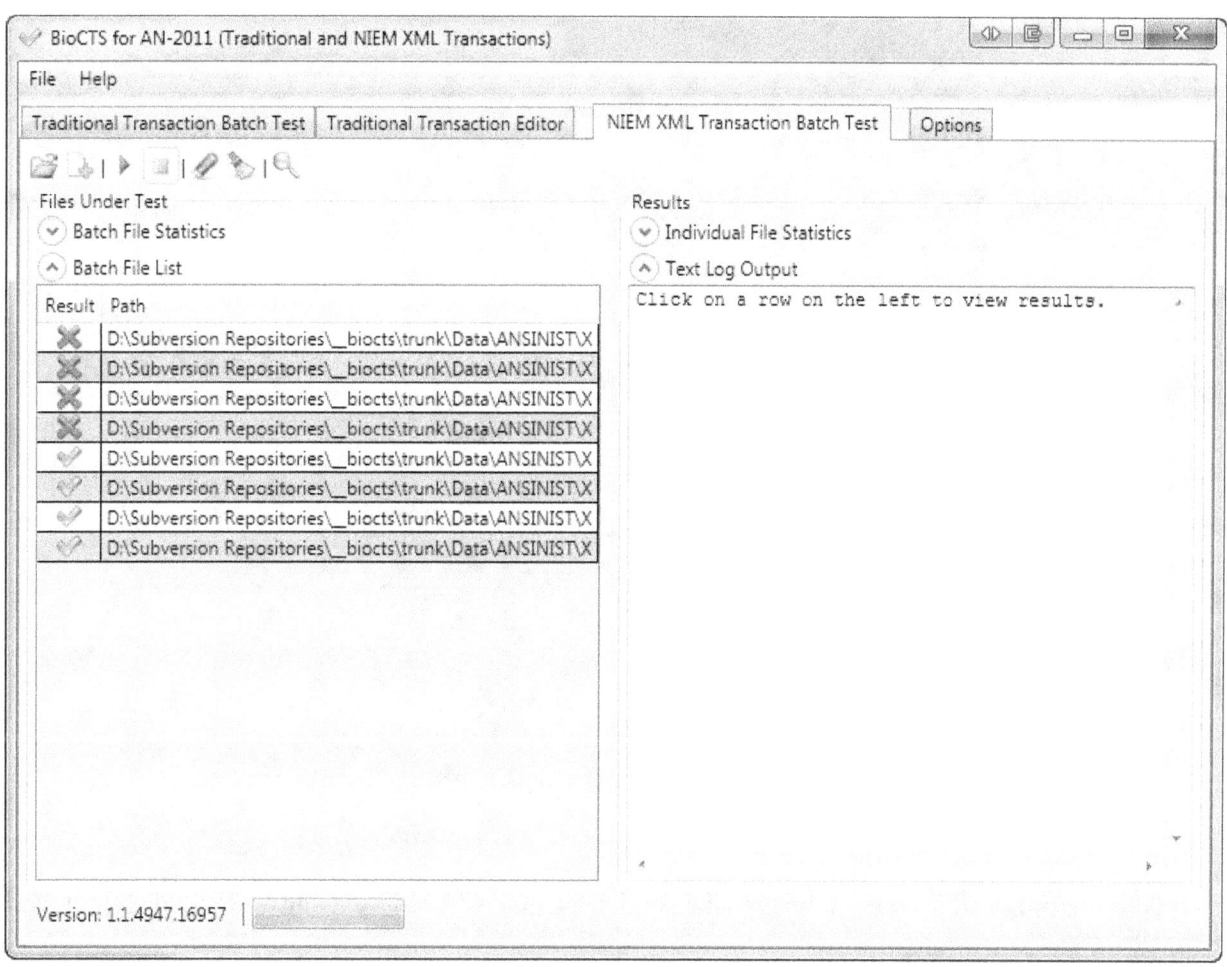

Figure 3-4 – Group of Files Loaded and tested - Overall Results Shown

Text output results for each transaction can be viewed by clicking on the desired filename in the "File Under Test" pane. Fig 3-5 shows how the complete text results are displayed.

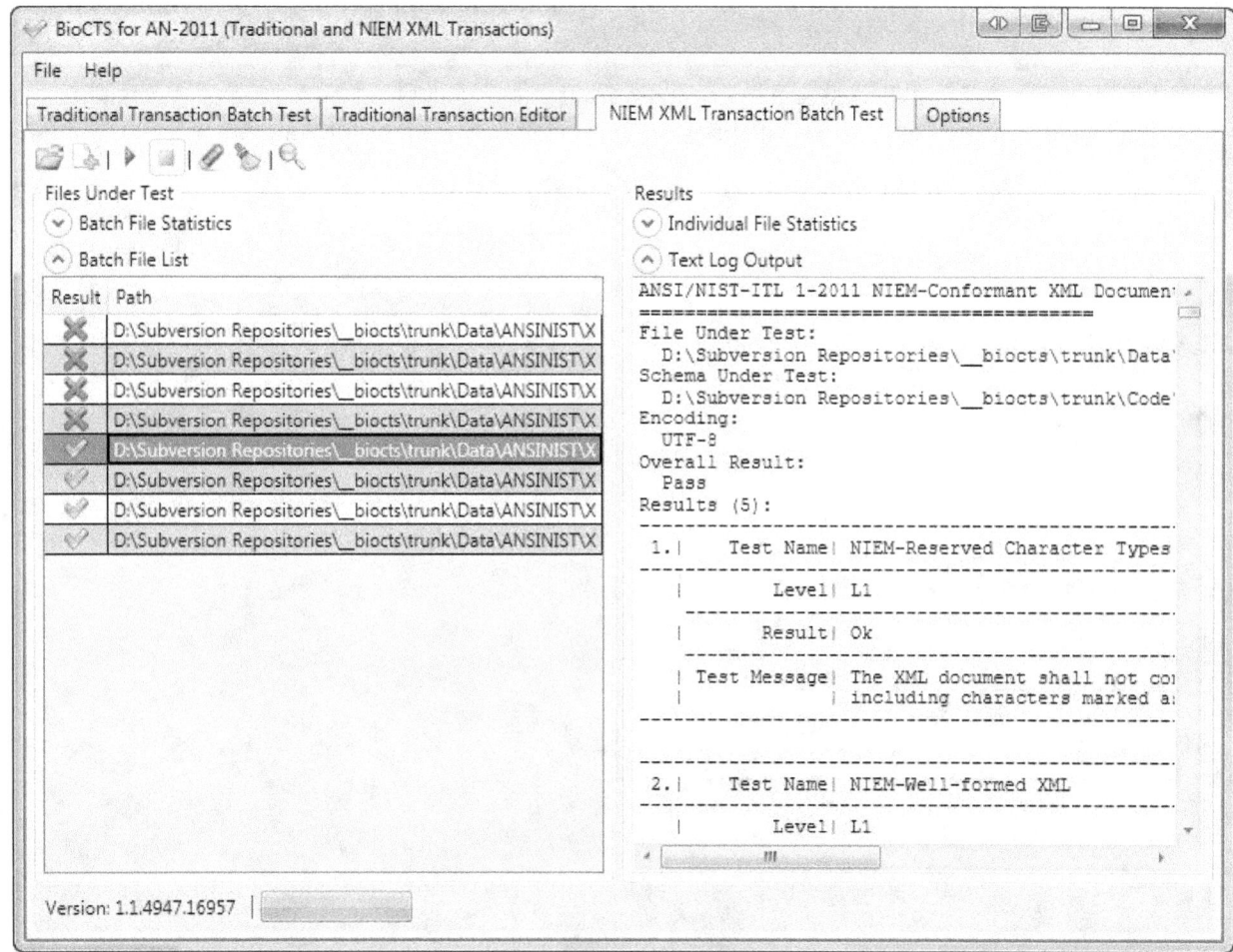

Figure 3-5 – Display of the Text Log Output of a File Tested

3.2.2 Options

Options are provided for output file type, constraint schema path and Text and XML Log Output save locations.

- *Output File Type*

BioCTS for AN-2011 is capable of two types of outputs:

- o Text Log Output - which is always generated during batch testing (for Traditional and NIEM XML encoded Transactions)

- o Optional XML Output, which will generate an XML Output log that includes the same information found in the Text Log.

Because of the amount of details provided in the Log Outputs, the size of these logs can be large.

- *Time-stamped Folder Format*

Output is generated in a Time-stamped named folder located in the output directory. The time-stamped folder named format is as shown below, with elements separated by a ".":

- o yyyy – 4 digit year (e.g. 2012)
- o MM – 2 digit month (e.g. 10)
- o dd – 2 digit day (e.g. 31)
- o HH – 2 digit hour in 24-hour scale (e.g. 13)
- o mm – 2 digit minutes (e.g. 59)
- o ss – 2 digit seconds (e.g. 22)

Examples:
- o Text Output will be generated in the directory:
  C:\Users\dyaga\Desktop\BioCTS for AN-2011 Output\2012.10.31.13.59.22\Text Output
- o XML Output will be generated in the directory:
  C:\Users\dyaga\Desktop\BioCTS for AN-2011 Output\2012.10.31.13.59.22\XML Output

- *User-Defined Constraint Schema*

As shown in Fig. 3-6, in addition to the options above, there is an option to load in a User-Defined Constraint Schema for the XML Encoded Transaction Testing (as allowed in section C.3 of the ANSI/NIST-ITL 1-2011 standard).

Clicking the "Default" button will restore the original schema file provided with BioCTS for AN-2011 XML.

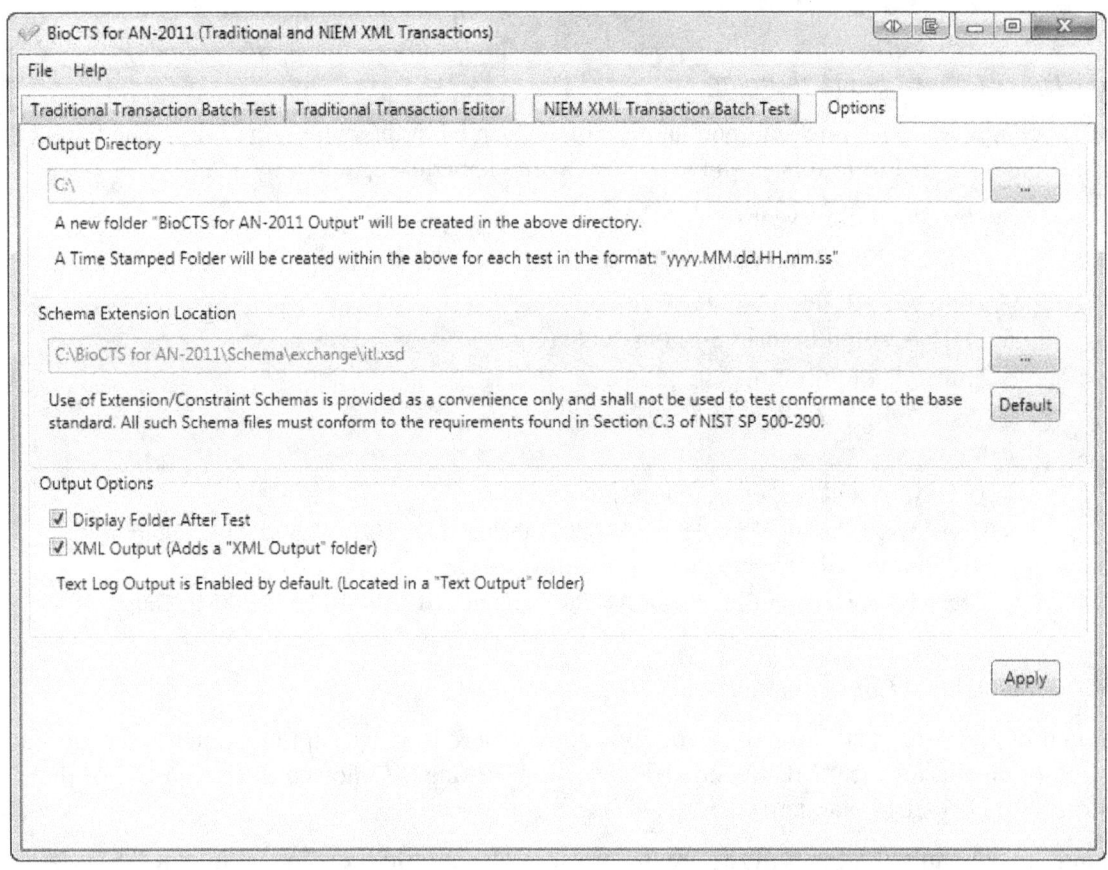

Figure 3-6 - The Options Tab with Schema Selection Highlighted

4 Result Log Detail

Every test performed in *BioCTS for AN-2011 NIEM XML* generates a Result – passing or failing. In an effort to enhance the readability and clarity of the Results, the textual output formatting has been modified since the initial release of *BioCTS for AN-2011*.

The Text Log Output Result format is shown in Fig. 4-1

```
:
-------------------------------------------------------------------------------------------------
1.|    Test Name| NIEM-Type1-1.001-First
-------------------------------------------------------------------------------------------------
   |       Level| L1
   -------------------------------------------------------------------------------------------------
   |      Result| Ok
   -------------------------------------------------------------------------------------------------
   | Test Message| Is the first element <biom:RecordCategoryCode>?
-------------------------------------------------------------------------------------------------
```

Figure 4-1 - A Text Log Result

The XML Log Output Result format is shown in Fig. 4-2.

```
<Result>
  <Level>L1</Level>
  <Message>The first element (biom:RecordCategoryCode) in the record shall be &lt;biom:RecordCategoryCode&gt;.</Message>
  <Results>Ok</Results>
  <Test>NIEM-Type1-1.001-First</Test>
</Result>
```

Figure 4-2 - An XML Log Result

A Result consists of, and clearly identifies:

- Test Name – The Test Names match up with what is found in NISTIR 500-295

- Test Level

  - Parse – A parse level test deals with any test that relates to parsing of the information; tests with this level happen before the data is inspected during L1, L2 and L3 testing.

  - L1 – A Level 1 test, which tests for values, lengths, and character counts of the data.

  - L2 – A Level 2 test, which tests for relationships between Fields, Subfields, Information Items and Records.

  - L3 – A Level 3 test, which tests to see if the data specified is consistent with the biometric sample presented.

- Test Result

  - Ok – The test was unable to find an error. This does not necessarily mean that this portion of data was without error; just that the tests could not find error.

25

- o Message – The test was unable to find an error; however the test found it necessary to convey an additional message.

- o Warning – The test was unable to find an error; however there may be an aspect that warrants further investigation.

- o Error – The test was able to find an error.

- o Critical – The test was able to find an error; this error was critical enough that it may impede further testing.

- Test Message – The message can contain any additional information to clarify the test.

5 References

The list below provides references for this publication.

[1] ANSI/NIST-ITL 1-2011, NIST Special Publication 500-290 Data Format for the Interchange of Fingerprint, Facial & Other Biometric Information, November 2011.

   http://www.nist.gov/customcf/get_pdf.cfm?pub_id=910136 (accessed 7/01/13).

[2] National Information Exchange Model Naming and Design Rules, NIEM Technical Architecture Committee (NTAC), October 31, 2008, Version 1.3.
   https://www.niem.gov/documentsdb/Documents/Technical/NIEM-NDR-1-3.pdf

[3] ISO 12052:2006 Health informatics – Digital imaging and communication in medicine (DICOM) including workflow and data management. Available at http://www.iso.org/iso/home/store.htm

[4] ANSI/NIST-ITL Standard Web page http://www.nist.gov/itl/iad/ig/ansi_standard.cfm (accessed 7/01/13).

[5] International Committee for Information Technology Standards Technical Committee M1 (INCITS/M1) Public Web Page: http://standards.incits.org/a/public/group/m1

[6] ISO/IEC JTC 1/SC 37 Web page http://www.iso.org/iso/jtc1_sc37_home

[7] Grother, P., Salamon, W., Chandramouli, R. NIST Special Publication 800-76-2, Biometric Specifications for Personal Identity Verification, July 2013

[8] BioCTS - NIST/ITL CSD Biometric Conformance Test Software for Biometric Data Interchange Formats Web Site: http://www.nist.gov/itl/csd/biometrics/biocta_download.cfm#CTAdownloads

[9] NIST/ITL's Biometric Application Programming Interface (BioAPI) Conformance Test Suite (CTS). Available at: http://www.nist.gov/itl/csd/biometrics/bioapicts.cfm

[10] NIST/ITL Conformance Test Suite for Patron Format A Data Structures Specified in ANSI INCITS 398-2008, Common Biometric Exchange Formats Framework (CBEFF). Available at: http://www.nist.gov/itl/csd/biometrics/biocbeffcts.cfm

[11] ANSI/NIST-ITL 1-2011 XML Exchange Package associated Schema, July 7, 2013.
   http://biometrics.nist.gov/cs_links/standard/ansi_2011/ANSi-Files.zip

[12] McGinnis, C. J., Yaga, D., and Podio, F. L., NIST IR 7806, ANSI/NIST-ITL 1-2011 Requirements and Conformance Test Assertions, September 2011.

[13] Podio, F. L., Yaga, D., and McGinnis, C. J., Editors, NIST Special Publication 500-295, Conformance Testing Methodology for ANSI/NIST-ITL 1-2011, Data Format for the Interchange of Fingerprint, Facial & Other Biometric Information (2012).

[14] "Well-formed document" Wikipedia: The Free Encyclopedia. Wikimedia Foundation, Inc., 23 April 2013. Web. 30 May 2013. http://en.wikipedia.org/wiki/Well-formed_document

[15] "Extensible Markup Language (XML) 1.0" World Wide Web Consortium (W3C), 10 February 1998. Web. 30 May 2013. http://www.w3.org/TR/1998/REC-xml-19980210#dt-wellformed.

[16] ANSI/NIST-ITL 1-2011 Sample Data Files, including schemas, sample traditional and XML transactions. http://biometrics.nist.gov/cs_links/standard/ansi_2011/ANSi-Files.zip

[17] W3C Recommendation 26 November 2008, Extensible Markup Language (XML) 1.0 (Fifth Edition), http://www.w3.org/TR/REC-xml/

Appendix A – Test Assertions Not Implemented in the Test Tool

There are requirements specified in the AN-2011 standard that do not have test assertions and are not implemented within the software. Table A-1 identifies and provides justification for all exceptions.

Exception	Section	AN-2011 Requirement Summary	Justification
Domain Names / Application Profile Specifications	5.3.2	Data contained in this record shall conform in format and content to the specifications of the domain name(s) as listed in Field 1.013 Domain name / DOM found in the Type-1 record, if that field is in the transaction. The default domain is NORAM. Field 1.016 Application profile specifications / APS allows the user to indicate conformance to multiple specifications. If Field 1.016 is specified, the Type-2 record must conform to each of the application profiles. A DOM or APS reference uniquely identifies data contents and formats. Each domain and application profile shall have a point of contact responsible for maintaining this list. The contact shall serve as a registrar and maintain a repository including documentation for all of its common and user-specific Type-2 data fields. As additional fields are required by specific agencies for their own applications, new fields and definitions may be registered and reserved to have a specific meaning. When this occurs, the domain or application profile registrar is responsible for registering a single definition for each number used by different members of the domain or application profile.	The format and content of the record are defined by the DOM or APS. Each DOM and APS has related record-content definitions that may be updated. The evolving nature of the DOM and APS definitions and nature of using registrars means that the requirements are not defined in the base standard.[3]
	6	An implementation domain, coded in Field 1.013 Domain name / DOM of a Type-1 record as an optional field, is a group of agencies or organizations that have agreed to use preassigned data fields with specific meanings (typically in Record Type-2) for exchanging information unique to their installations. The implementation domain is usually understood to be the primary application profile of the standard. New to this version of the standard, Field 1.016 Application profile specifications / APS allows multiple application profiles to be referenced. The organization responsible for the profile, the profile name and its version are all mandatory for each application profile specified. A transaction must conform to each profile that is included in this field. It is possible to use Field 1.016 and / or Field 1.013. A specified implementation domain and specified application	Since the "transaction must conform to each profile" included in the field, and those profiles are defined by the listed agency, the CTS would have to retrieve the latest requirements from the agency.[1]

---

[3] Requirements related to user, profile, or domain-specific information are not within the scope of conformance testing to the base AN-2011 standard, and therefore are not tested by the BioCTS for AN-2011 software developed to test AN-2011 Traditional and NIEM XML Encoded Transactions..

		profiles must all have the same definition for fields, subfields and information items that are contained in the transaction.	
Alternate Character Sets	5.6, Table 2	Field 1.015 Character encoding/DCS is an optional field that allows the user to specify an alternate character encoding… Field 1.015 Character encoding/DCS contains three information items: the character encoding set index/ CSI, the character encoding sent name/CSN, and the character encoding set version/CSV. The first two items are selected from the appropriate columns of Table 2.	Table 2 lists ASCII, UTF-16, UTF-8, and UTF-32 as possible encodings. However, the table also allows "User-defined" character encoding sets. [1]
Alternate Coordinate System	7.7.3, Table 4	The ninth information item is the geodetic datum code / GDC10. It is an alphanumeric value of 3 to 6 characters in length. This information item is used to indicate which coordinate system was used to represent the values in information items 2 through 7. If no entry is made in this information item, then the basis for the values entered in the first eight information items shall be WGS84, the code for the *World Geodetic Survey 1984 version - WGS 84 (G873)*. See Table 4 for values.	Table 4 lists 22 coordinate systems and the option to include "Other" types as well. It is not feasible for the CTS to test conformance to all coordinate systems, specifically those that are listed by the user under "Other". [1] The CTS tests for conformance to WGS84 because it is the default coordinate system used in the base standard.
	7.7.3	A fourteenth optional information item geographic coordinate other system identifier / OSI allows for other coordinate systems. This information items specifies the system identifier. It is up to 10 characters in length. Examples are: • MGRS (Military Grid Reference System) • USNG (United States National Grid) • GARS (Global Area Reference System) • GEOREF (World Geographic Reference) • LANDMARK (e.g. hydrant) and position relative to the landmark. A fifteenth optional information item, is the geographic coordinate other system value / OCV. It shall only be present if OSI is present in the record. It can be up to 126 characters in length. If OSI is LANDMARK, OCV is free text and may be up to 126 characters. For details on the formatting of OCV for the other coordinate systems shown in OSI as examples, see http://earth-info nga.mil/GandG/coordsys/grids/referencesys.html	While some examples of other coordinate systems are listed in the standard (MGRS, USNG, GARS, GEOREF, LANDMARK), those values are not all-inclusive, and the user may indicate other coordinate systems that are not listed[1].
Subject Acquisition Profiles SAP/FAP/I	7.7.5, Table 8, Table 9, Table 10	A subject acquisition profile is used to describe a set of characteristics concerning the capture of the biometric sample. These profiles have mnemonics SAP for face, FAP for	It is not feasible to test if the image was captured under the conditions specified by the SAP, FAP

AP		fingerprints and IAP for iris records.	or IAP level as defined in Tables 10 through 13. However, the fields will be tested for valid level values.
Open and Closed Paths	7.8	Several Record Types define open paths (also called contours or polylines) and / or closed paths (polygons) on an image. They are comprised of a set of vertices. For each, the order of the vertices shall be in their consecutive order along the length of the path, either clockwise or counterclockwise. (A straight line of only two points may start at either end). A path may not have any sides crossing. No two vertices shall occupy the same position. There may be up to 99 vertices. An open path is a series of connected line segments that do not close or overlap. A closed path (polygon) completes a circuit. The closed path side defined by the last vertex and the first vertex shall complete the polygon. A polygon shall have at least 3 vertices. The contours in Record Type-17: Iris image record can be a circle or ellipse. A circle only requires 2 points to define it (See Table 16). There are two different approaches to the paths in this standard. The 2007 and 2008 version of the standard used paths for Field 14.025: Alternate finger segment position(s) / ASEG. That approach has been retained in this version for all paths except in the Extended Feature Set (EFS) of Record Type-9. The EFS adopted an approach expressing the path in a single information item, which is different than that used in other record types.	Further research is needed to determine the feasibility of testing for: -simple, plane figure -no sides crossing -no interior holes

Table A-1 Test Assertions Not Implemented in the Test Tool

Appendix B - Changes Made to the Schema Files

Below are the changes that have been made to the schema files as of the time of writing.

Additions and Modifications to: *subset\niem\domains\biometrics\1.0\biometrics.xsd*

- **Added**: Values (128 to 999) to the Simple Type *CharacterSetIndexCodeSimpleType*
  Reason: The values listed in the original schema prevented the use of "User-defined character encoding sets" as specified in Table 4 of the AN-2011 standard which are specified as values 128 to 999.

- **Added**: A *minOccurs="0"* to the element *biom:FaceImageAcquisitionProfile* of the complex type *FaceImageType*.
  Reason: As specified in Table 57 of the AN-2011 standard, Field 10.013 has a min occurrence of 0, max occurrence of 1. By not having the *minOccurs="0"* the schema was requiring this element to always be present.

- **Added**: The value 18 for "Unknown friction ridge" to the Simple Type *FingerPositionCodeSimpleType*.
  Reason: Value 18 is specified as a valid value in Table 8 of the AN-2011 standard, but was not present in the schema.

- **Added**: A new simple type: *AlphabeticStringSimpleType* which allows the alphabetic characters and space (characters *[a-zA-Z\s]*), for zero or more times.
  Reason: It was used as a basis for the following type:

  - **Added**: A new Type *OneToSixteenCharacterAlphabeticStringSimpleType* – This puts a min length of 1 and max length of 16 to the *AlphabeticStringSimpleType*
    Reason: This new type was used to amend the *TransactionCategoryCode* which specifies a list of enumerated values. Just having the list of values was preventing the base requirement of having a user-defined field of Alphabetic strings 1 to 16 characters long (Field 1.004 in Table 22 of the AN-2011 standard)

- **Modified**: The Element *FingerprintImageStitchedIndicator* to only accept the character "Y".
  Reason: The Element was defined in the schema as a Boolean type (*true* or *false*). This prevented the base requirement of only allowing a single alphabetic character of "Y" (Field 14.027 in Table 71 of the AN-2011 standard).

www.ingramcontent.com/pod-product-compliance
Lightning Source LLC
Chambersburg PA
CBHW081407170526

45166CB00010B/3238